Terrific Trucks!

GARBAGE TRUCKS

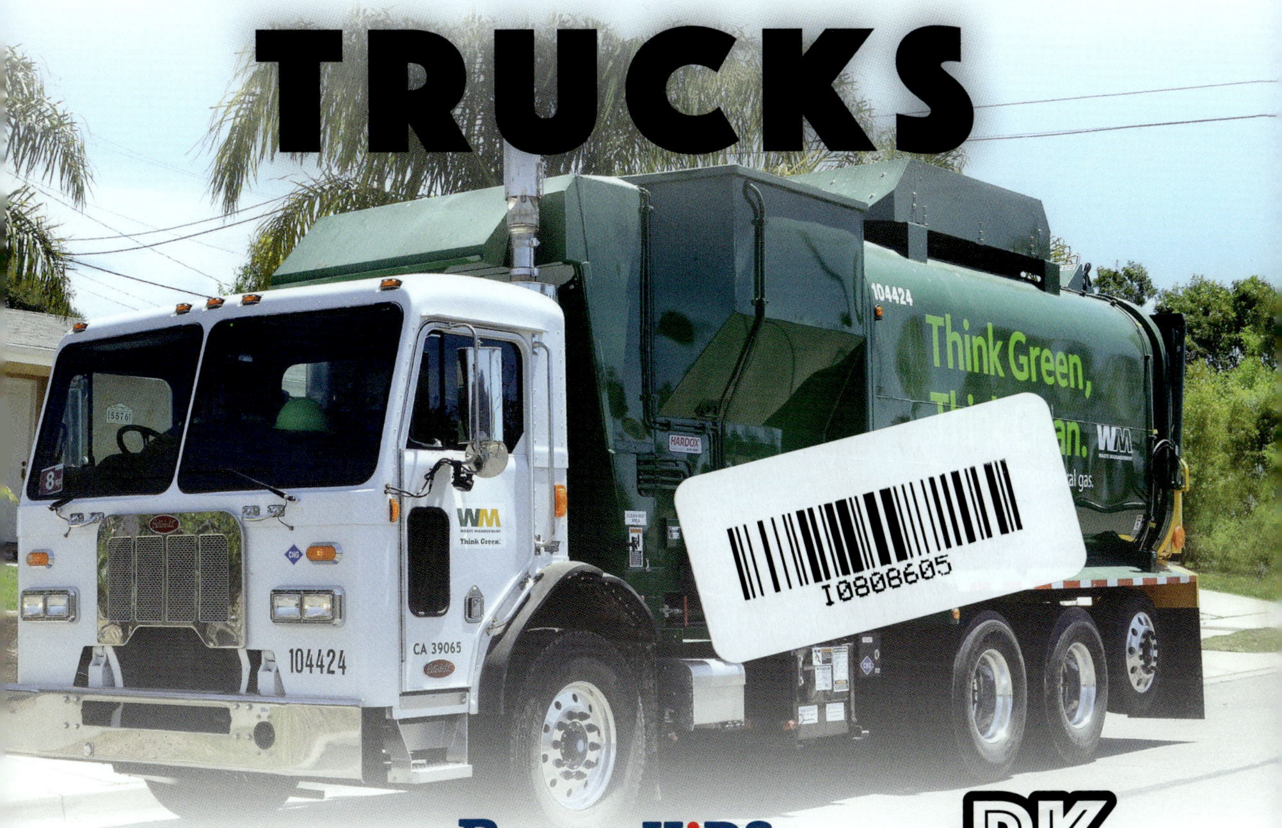

Ursula Pang

PowerKiDS press

PK Beginners

Here comes
the garbage truck!

sanitation
25DP-440
Bk.S 10
SAFETY
SERVICE
SUSTAINABILITY
25DP-440
Bk.10
NEW YORK
City of
New York

The garbage truck
is big.

2454
Companies driven
to save lives
MACK
2454
TYNGSBORO, MA
USDOT 1011363
2454
REPUBLIC
SERVICES

The garbage truck
is loud.

A worker drives
the garbage truck.

The garbage truck stops at our house.

2471
REPUBLIC
SERVICES
republicservices.com
CAUTION
VEHICLE STOPS AND BACKS FREQUENTLY

A worker puts garbage in the truck.

Rings & More

The garbage truck goes to the next house.

100

The garbage truck
is getting full.

The garbage truck makes the garbage flat.

The garbage truck
takes garbage
to the garbage dump.

Good bye,
garbage truck!